RAND SOCIAL AND ECONOMIC WELL-BEING

Community Citizen Science

From Promise to Action

Ramya Chari, Marjory S. Blumenthal, Luke J. Matthews

For more information on this publication, visit www.rand.org/t/RR2763

Library of Congress Cataloging-in-Publication Data is available for this publication.
ISBN: 978-1-9774-0306-3

Published by the RAND Corporation, Santa Monica, Calif.
© Copyright 2019 RAND Corporation
RAND® is a registered trademark.

Cover: VectorMine/Adobe Stock.

Support RAND
Make a tax-deductible charitable contribution at
www.rand.org/giving/contribute

www.rand.org

Preface

Citizen science is the use of scientific methods by the general public to answer questions about the world and solve problems of concern. Today, citizen science is used for numerous activities, such as monitoring ecosystem health, enhancing disaster preparedness, and informing clinical research. These examples illustrate not only the breadth of the field but also its dual applications for supporting scientific advancements and informing decisionmaking. Currently, there is a lack of research on models, such as community citizen science, that focus on using data to inform policy or community actions. Assessments of challenges and opportunities for citizen science implementation are needed to maximize its potential for impact. Recognizing the need for foundational research in this rapidly evolving area, RAND Ventures funded a two-year project to explore citizen science and its potential for research and policy applications. This report, intended for a general audience, builds on earlier work published in 2017, *The Promise of Community Citizen Science* (Chari et al., 2017), and provides a greater understanding of how community citizen science can best be translated into real-world actions.

RAND Ventures

The RAND Corporation is a research organization that develops solutions to public policy challenges to help make communities throughout the world safer and more secure, healthier and more prosperous. RAND is nonprofit, nonpartisan, and committed to the public interest.

RAND Ventures is a vehicle for investing in policy solutions. Philanthropic contributions support our ability to take the long view, tackle tough and often-controversial topics, and share our findings in innovative and compelling ways. RAND's research findings and recommendations are based on data and evidence, and therefore do not necessarily reflect the policy preferences or interests of its clients, donors, or supporters.

Funding for this venture was provided by gifts from RAND supporters and income from operations.

RAND Community Health and Environmental Policy Program

RAND Social and Economic Well-Being is a division of the RAND Corporation that seeks to actively improve the health and social and economic well-being of populations and communities throughout the world. This research was conducted in a precursor to the Community Health and Environmental Policy Program within RAND Social and Economic Well-Being.

The program focuses on such topics as infrastructure, science and technology, community design, community health promotion, migration and population dynamics, transportation, energy, and climate and the environment, as well as other policy concerns that are influenced by the natural and built environment, technology, and community organizations and institutions that affect well-being. For more information, email chep@rand.org.

Tables

Summary

Citizen science is the use of scientific methods by the general public to ask and answer questions about the world and solve problems of concern. The field is growing rapidly. With new advancements in digital technologies, avenues for collaboration, invention, and data collection are now placed at the fingertips of millions (Bishop, 2014; Newman et al., 2017). Today, there are unprecedented opportunities for public engagement in science. For example, citizen science activities are taking place to monitor ecosystem health, enhance disaster preparedness, inform clinical research, make astronomical discoveries, and measure environmental health hazards. These examples illustrate the breadth of the field and its relevance for supporting scientific advancements and informing decisionmaking.

Although uses of citizen science are proliferating, there has been no robust assessment of the challenges and opportunities associated with translation of citizen science research into actions. Our prior work explored a type of citizen science in which citizens exert a high degree of control over research (Chari et al., 2017). This model encompasses forms of citizen science in which citizens work together with professional scientists throughout the research process and in which citizens perform research on their own (Shirk et al., 2012). Building on existing typologies by such scholars as Candie C. Wilderman (Wilderman, 2007), Rick Bonney and his colleagues (Bonney et al., 2009), and Jennifer L. Shirk and her colleagues (Shirk et al., 2012), we call this model *community citizen science*. Given its promise in empowering communities to perform scientific research, community citizen science is an important, yet understudied, model. Also, community citizen science activities often focus on addressing community concerns; accordingly, a better characterization could yield insights into important barriers and opportunities for translating community citizen science research into action.

In this report, we aimed to (1) characterize the nature of community citizen science and its potential uses, (2) identify implementation needs and challenges, (3) conceptualize pathways through which community citizen science can achieve policy and community impacts, and (4) elucidate challenges that community citizen science might face in reaching its goals. To achieve these objectives, we performed a qualitative analysis consisting of 30 semistructured interviews with citizen science experts from the academic, government, community, and private sectors. Five of these interviews were with project leaders representing three community citizen science projects carried out for a specific application: disaster response and recovery. These case studies were

- SkyTruth pollution-tracking applications (SkyTruth, undated b, undated c)
- Planetary Response Network activations for disaster response (Zooniverse, undated a, undated b)

- the Surfrider Foundation's Rincón chapter's Blue Water Task Force and its Hurricane Maria activities (Dias, 2017).

These projects were deployed in recent disasters, including Hurricanes Harvey, Irma, and Maria in 2017; the 2016 earthquake in Ecuador; and the 2010 *Deepwater Horizon* oil spill.

We performed thematic analyses that integrated perspectives from both community citizen science leaders and experts across the larger field of citizen science. This approach allowed us to understand how contextual elements could influence implementation and outcomes of community citizen science activities.

Findings

We used findings from our disaster response and recovery case studies to identify specific considerations for community citizen science implementation and impacts; interviews with general citizen science experts provided a greater understanding of how the cultures and institutions that make up the larger field of citizen science can shape and react to the development of community citizen science. In this section, we summarize the major themes from our interviews.

Community Citizen Science and Traditional Scientific Institutions Reflect Different Cultures

Community citizen science (also known as grassroots science) models enjoyed general support from all interviewees; however, it appeared that the further away citizen science moved from the trappings and traditions of *academic research* (i.e., professional or academic-led citizen science), the more questions arose about the ability of community citizen scientists to utilize sound scientific methods and the proper role of traditional scientists within these forms. From the community citizen science perspective, there was awareness of the scrutiny that projects might face and therefore a desire to adhere to standards for performing quality research. To perform their research, however, citizens leading community citizen science projects might not be replicating academic research models. Rather than being characterized by academic degrees, research scholarship, and institutional affiliations, these *citizen research* models might instead be based on such elements as action-based research, provision of technical services, and a nontraditional mix of institutions engaged as research partners (e.g., advocacy groups, local businesses, service organizations). Both academic and citizen research models share a common goal of performing "good science," but differences in culture, goals, activities, and actors can have ramifications for resource needs, implementation activities, and the acceptance and use of research. Even the more collaborative models, in which citizen scientists work closely with traditional scientists, could raise concerns in larger academic circles about rigor and quality. Unfamiliarity with citizen science appears to be a driver of such concerns, but interviewees also noted a need for scientific and policymaking institutions to change their cultures to promote greater acceptance of citizen science, allow for sharing of knowledge and resources, and incentivize professional scientists to pursue collaborative research.

Community Citizen Science Can Creatively Leverage Resources to Address Project Implementation Challenges

Implementation of citizen science projects in general appears to face the same kinds of challenges regardless of whether citizen scientists are leading activities or just participating in research. These challenges include

- scaling and replicating citizen science activities
- balancing the roles of professional and citizen scientists
- navigating competition and duplication within the field
- clearly communicating project mission and objectives to volunteers
- obtaining needed resources, such as funding, technology and infrastructure, partnerships, volunteer participation, organizational support, and committed and capable leadership.

In addition, disaster-related projects might exhibit their own set of challenges related to such issues as volunteer engagement and recruitment during chaotic times or changing circumstances on the ground that affect project planning and management.

Although challenges might be similar, the ways in which community citizen science projects and other types of citizen science activities address these challenges might differ. For example, case study interviewees described a greater range of partner types and highlighted the importance of relationships with entities not traditionally involved in scientific research, such as media groups, local businesses, and service organizations. These types of partnerships could help offset the resource limitations under which community citizen science projects sometimes operate and could be particularly useful in disaster settings if traditional scientific infrastructure is disrupted.

Conveying Credibility and Promoting Community Citizen Science Research Will Be Important for Meeting Policy-Related Objectives

Interview participants noted that, across the field of citizen science as a whole, there is potential for many scientific and social benefits, including

- improved research processes and outcomes
- enhanced educational opportunities and scientific literacy
- greater individual and community problem-solving abilities and self-efficacy beliefs
- increased public engagement in scientific and civic endeavors
- stronger community networks.

However, unlike general citizen science interviewees, who tended to describe the purpose of projects in terms of achieving scientific or educational goals, community citizen science interviewees characterized their work as a reaction to scientific or policymaking institutions not performing (or being unable to perform) perceived or actual real-world responsibilities. Therefore, community citizen scientists were motivated to augment or extend research capabilities of scientific or policymaking institutions, undertake research functions that institutions do not or cannot perform, monitor institutional activities to promote accountability, or raise awareness and provide information for interventional efforts. We note, however, that this contrast might be the result of our focus on disaster-related community citizen science in particular, which is, by its nature, directly applied to events on the ground.

Given the focus of disaster community citizen science on informing preparedness, response, or recovery efforts, interviewees considered conveying the credibility of both researchers and data as highly important. To this end, interviewees discussed ways to address issues related to data quality, when community citizen science research might be most appropriate, how research activities and findings could best be communicated, and the importance of robust evaluation for continual improvement.

Conclusions and Future Directions

Our findings indicate that community citizen science is a promising means for conducting applied research and informing policies, particularly for disaster preparedness. The literature presents examples of applied community citizen science beyond just disasters, in such areas as environmental health (Public Lab, undated) and conservation (Hopkins and Freckleton, 2002). Our case studies demonstrated that community citizen science for disaster preparedness can be valuable in supporting official preparedness actions and that, rather than being a limiting factor, working outside traditional research institutions and infrastructure can lead to creative means of finding resources and bringing together diverse partnerships. According to our assessment, one of the major challenges to community citizen science could lie in changing existing scientific and policy cultures to accept and foster community citizen science.

Guidance for developing and implementing community citizen science projects is limited, and interviewees provided valuable advice on how to address resource challenges in such areas as partnerships and funding, technology and infrastructure, volunteer participation, and organizational support and leadership. However, research is needed to more systematically and comprehensively identify practices that might be relevant for community citizen science applications beyond disaster response and recovery. In addition, to move the field of community citizen science forward, more research is needed in general on the structures, outcomes, impacts, and trends exhibited by models of community citizen science that might be developing and operating outside of traditional scientific institutions.

Overall, community citizen science is a growing and dynamic area filled with unique opportunities for public engagement in science and the enhancement of community civic life. However, although the field is filled with promise, our work also revealed challenges facing its development and translation of community citizen science research. Ultimately, regardless of whether science is carried out by professionals or amateurs, in academic labs or community backyards, what matters is a shared commitment by all to the conduct of "good science." With that shared commitment, community citizen science can be a mechanism for bringing society together, under a common goal of pursuing knowledge to improve the health and well-being of all.

Acknowledgments

We would like to acknowledge Anne Bowser, director of innovation at the Woodrow Wilson International Center for Scholars, and Lori Uscher-Pines, senior policy researcher at RAND, for their comprehensive and insightful reviews of this report and their excellent comments and recommendations that served only to strengthen the final product. We are also grateful for the time and valuable perspectives freely offered by our interview participants. Our interviewees are exceptional leaders of the rapidly evolving world of citizen science, and our work stands on the shoulders of these pioneers. In addition, we would like to thank Amanda F. Edelman and Therese Jones, both of RAND, for their contributions in conducting interviews and performing analyses. Finally, we would like to acknowledge RAND leadership and colleagues involved in the administration of the RAND-sponsored research program and thank them for providing us with the opportunity to contribute to the development of this important field of policy research. In particular, we appreciate the steady guidance and exhortations provided by Howard J. Shatz as our work progressed.

Abbreviations

BWTF   Blue Water Task Force

CBPR   community-based participatory research

PRN   Planetary Response Network

CHAPTER ONE

Introduction

Citizen science is the use of scientific methods by the general public to ask and answer questions about the world and solve problems of concern. At once a new and old phenomenon, citizen science is rooted in the rich history of amateur science across the world that has generated important discoveries and shaped the course of many lines of scientific inquiry (Browne, 1995; Leavitt, 2010; van Wyhe, undated). Charles Darwin and Gregor Mendel were both seminal figures who greatly expanded knowledge of the natural world. And they were also *citizen scientists*—nonprofessionals who freely chose to study evolution and genetics, respectively, at times outside of the scientific institutions of their day. In more-modern times, citizen science has had a long tradition in the ecological and environmental sciences. With new advances in digital and mobile technologies, use of citizen science is being explored by a broader range of disciplines inside and outside of academia, and the field is growing rapidly. Importantly, modern-day technologies provide avenues for collaboration, sharing, invention, and data collection right at the fingertips of millions (Bishop, 2014; Newman et al., 2017). Today, there are unprecedented opportunities for public engagement in science. For example, citizen scientists are monitoring ecosystem health; generating astronomical discoveries; enhancing disaster preparedness, response, and recovery; measuring environmental health hazards; and informing clinical research. These examples illustrate the breadth of the field and its relevance for supporting scientific advancements and informing decisionmaking.

Citizen science does not differ from traditional science in terms of its range or breadth of uses. However, the potential ability of citizen science to expand research capabilities through increased labor effort, data volumes, or diversity of perspectives raises implications for the use of science in policy and societal applications. Although uses of citizen science are proliferating, to date, there has been little focus on understanding the challenges and opportunities that arise when translating citizen science research into decisions or actions. Existing research, focused on the policy relevance of citizen science, tends to examine policy and decisionmaking outcomes in cases in which citizen science activities are led by academic, government, or other institutional actors (Haklay, 2015). What is missing is an emphasis on the *citizen* aspects of citizen science and the perspectives of the nonprofessional or amateur scientists who are on the ground collecting and analyzing data.

What Is Community Citizen Science?

Many scholars charting the modern citizen science movement have characterized activities according to the level of public involvement and degree to which citizens exhibit control or ownership over the work. For example, Candie Wilderman categorized citizen science based on answers to five questions: (1) Who defines the problem? (2) Who designs the study? (3) Who collects the samples? (4) Who analyzes the samples? and (5) Who interprets the data? (Wilderman, 2007). Informed by the answers, she developed three models: *community consulting*, in which communities ask experts for help with a specific question (science for the people); *community workers*, in which the public is involved in data collection and some analyses for studies run and designed by experts (science with the people); and *community-based participatory research* (CBPR), in which communities are taught to lead through all aspects of the research, from problem definition to data interpretation (science by the people). Other researchers have proposed similar frameworks that differ in the parsing of specific categories. For example, Rick Bonney and his colleagues called Wilderman's community-worker model *contributory* and the CBPR model *cocreated* (the public is involved in the full scope of the research endeavor, from defining the question to disseminating the results) (Bonney et al., 2009). They also defined a third category, *collaborative*, in which the public is involved in some aspects of the research, including developing data-collection protocols and hypotheses and interpreting the data to reach conclusions. Jennifer Shirk and her colleagues proposed two additional categories to the Bonney et al., 2009, framework: *contractual* activities (akin to the community-consulting model in Wilderman et al., 2007) and *collegial* activities (in which non-experts conduct independent investigations) (Shirk et al., 2012).

Our discussion on community citizen science situates itself at the level at which citizens exert control and ownership in scientific research beyond data collection and up to and including outright independence from established experts (see Chari et al., 2017). Using the Shirk et al., 2012, framework, our report focuses on the collaborative, cocreated, and collegial models of citizen science. These models have been given other names in the literature, including *community science*, *civic science*, and *street science*. Given the lack of a standard nomenclature (Eitzel et al., 2017), we refer to these models collectively as *community citizen science*.

Table 1.1 builds on the Shirk et al., 2012, citizen science typology to summarize the governance structures of different models, including three models that encompass our definition of *community citizen science*. The current citizen science literature has a larger focus, on contributory forms of citizen science (Dickinson et al., 2012; Smith et al., 2017), and a rich body of research exists on participatory research methods, such as CBPR or participatory action research, that provide insight into the strengths and weaknesses of collaborative and cocreated models (Freeman et al., 2006; Cargo and Mercer, 2008; Ross et al., 2010). Collegial models, however, are understudied in the existing literature. Table 1.1 illustrates that the defining feature of our conceptualization of community citizen science is (1) "meaningful" citizen engagement in aspects of the research process beyond data collection or (2) citizens in research leadership positions (and, of course, activity with both of these features would fall under our definition as well). We also note that citizen scientists can be both research leaders and research volunteers, but these are not necessarily the same people. For example, in a collegial research model, citizen scientists can lead projects that use contributory or crowdsourcing methods, involving volunteers mainly in data-collection activities. Therefore, we would categorize any research that is citizen led as collegial, regardless of whether research was conducted in a contributory or collaborative fashion.

In its idealized form, through deeply engaging the broader public in problem-solving, knowledge generation, and research translation, community citizen science could lead to more-robust, more-open, and more-democratic decisionmaking processes (Ottinger, 2009).

Table 1.1
Governance Structures of Citizen Science Models

Role	Model				
	Contractual	Contributory	Collaborative[a]	Cocreated[a]	Collegial[a]
Project initiator	Citizen	Academic	Academic	Citizen or joint	Citizen
Research lead	Academic	Academic	Academic	Academic	Citizen
Citizen[b]					
Leadership	No	No	No	Yes	Yes
Research	No	Yes	Yes	Yes	Yes
Possible citizen research					
Problem definition[c]	x		x	x	x
Study design			x	x	x
Data collection		x	x	x	x
Analysis			x	x	x
Interpretation			x		x

[a] Community citizen science model.

[b] Citizen leadership and citizen researchers are not necessarily the same people.

[c] In this table, *research* refers to actually carrying out research tasks. Problem definition is still a possible citizen research role, though: Citizens can identify the problem they need solved, but then scientists are the ones who actually define the problem.

The model therefore creates avenues for communities to foster education and literacy, build social capital, and grow future leaders. Community citizen science activities also often focus on addressing community concerns; accordingly, a better characterization of this model might yield insights into important barriers and opportunities for translating its research into action.

Report Goals

For this report, we had four goals:

- characterize the nature of community citizen science and its potential uses
- identify implementation needs and challenges
- conceptualize pathways through which community citizen science can achieve policy or community impacts
- elucidate challenges that community citizen science might face in reaching its goals.

To achieve these objectives, we performed a qualitative analysis consisting of 30 semistructured interviews with citizen science experts from the academic, government, community, and private sectors. Five of these interviews were with project leaders representing three community citizen science projects carried out for a specific application: disaster response and recovery (we call these our case studies). This report describes some of the first empirical research designed to understand the factors that might influence translation of citizen science research to action, with a particular focus on community citizen science. Through analysis and synthesis of expert perspectives, our goal is to provide information that can help citizen science researchers and practitioners effectively implement projects, address challenges, and achieve lasting impacts for the benefit of science and society.

CHAPTER TWO

Methods

Our qualitative approach consisted of 30 semistructured interviews with citizen science experts from the academic, government, community, and private sectors. Five of these interviews, which we call our case studies, were with project leaders representing three community citizen science projects that carried out disaster response and recovery. We conducted our interviews between April 2017 and June 2018. RAND's institutional review board approved our data-collection and management procedures.

Interview Participants and Sampling Strategy

General Citizen Science Experts

For general citizen science experts (non–case studies), we used maximum variation sampling (Coyne, 1997) to capture a range of perspectives across organizations (e.g., government, academic research, policy research) and scientific and technical disciplines (e.g., ecology, environmental science, public health, application development). We constructed a list of 35 potential interviewees identified from (1) peer-reviewed literature, (2) gray literature sources (e.g., white papers, magazine articles), (3) Citizen Science Association working group member lists available online, and (4) RAND team networks. Given the large share of citizen science activity involving government and academia and in the fields of ecology and environmental education, we identified greater numbers of potential interviewees in these categories. Our process was also driven by expert prominence. For example, we included known citizen science thought leaders regardless of organization or disciplinary category.

Out of 35 people contacted, 23 agreed to participate. We used snowball sampling (Patton, 2002) to identify five additional people, for a total of 40 potential interviewees. Of the five, two agreed to participate. The final general interview set consisted of 25 interviewees. Table 2.1 describes the interviewees' characteristics.

Community Citizen Science Case Study Participants

Given the range of possible community citizen science projects, we focused additional interviews on a common application, to limit the extent of variation across projects. Disaster response and recovery stood out as a promising application for several reasons. First, disaster-related citizen science projects, by their nature, already focus on using research to inform actions for response or recovery. Secondly, a large proportion of existing community citizen science projects focus on disasters in some form (e.g., hurricanes, climate change). Finally, disaster

Table 2.1
Interviewee Characteristics

Interviewee Type	Category		Number Contacted	Number Interviewed
General expert	Organization type	Government	10	7
		Academic or scientific research	11	5
		Policy research	3	3
		Community or nonprofit	4	1
		Education	5	4
		Technology development	4	2
		Private industry	3	3
	Discipline type	Astronomy	4	3
		Ecology	10	7
		Environmental science	4	3
		Environmental education	6	3
		Health and public health	4	2
		Policy-related area	3	2
		Technology-related area	9	5
	Total		40	25
Case study	Organization type	Community or nonprofit	4	4
		Technology development	1	1
	Citizen science model	Collaborative	1	1
		Collegial	4	4
	Total		5	5

situations are dynamic and inject unpredictability into project management. Therefore, these types of projects might offer insights into navigating both routine and exceptional challenges.

We first identified disaster events that occurred in the summer and fall of 2017, which, at the time of the study, represented the previous six months. The disaster pool included such events as hurricanes in the Caribbean, earthquakes in Mexico, wildfires across the western United States, and flooding in Africa and south Asia. We next conducted Google searches to find citizen science projects associated with each disaster category. To be eligible for selection as a case study, a project had to (1) be a collaborative, cocreated, or collegial citizen science model (see Table 1.1 in Chapter One) and (2) use research results to inform decisionmaking or community actions. Three projects met our criteria for community citizen science and were associated with major hurricanes in the fall of 2017 (Harvey, Irma, and Maria): projects carried out by SkyTruth (Harvey); the Planetary Response Network (PRN) (Irma and Maria); and the Rincón, Puerto Rico, chapter of the Surfrider Foundation (Maria). In Chapter Three, we describe the case studies in more detail.

We contacted representatives from the three projects, and all agreed to participate. Following interviews, we contacted additional people whom interviewees recommended for either formal interviews or targeted follow-up questions. In total, we interviewed five people and followed up with one person for greater clarity on specific issues (we exclude this follow-up interview from our interview totals). Four interviewees represented community or nonprofit organizations, and one represented technology or platform developers (see Table 2.1).

Data Collection

For all interviews, we developed a semistructured interview protocol that allowed us to explore interviewee perspectives. We used many of the same questions for both general expert and case study protocols but added project background and implementation questions for the case studies. Protocols consisted of the following topics: (1) citizen science benefits and impacts; (2) citizen science motivations, goals, and research uses; (3) implementation barriers, facilitators, and required resources; and (4) citizen science contexts, trends, and future directions.

We conducted the interviews by phone and recorded them with interviewee permission. Each lasted between 60 and 90 minutes. To supplement and confirm background details for the case studies, we reviewed publicly available documents from organization websites and gray literature.

Data Analysis

We combined the case study and general expert interviews into a single corpus for data analysis. We transcribed all interviews and coded them thematically using Dedoose version 8.0.35, a web application for managing, analyzing, and presenting qualitative and mixed-method research data (Dedoose, 2018). To define thematic areas, we first performed a deductive approach to coding using the four topics that defined the interview protocol (see "Data Collection"). We developed a hierarchically organized codebook to summarize themes and identify patterns across interviews using interviewee type (general expert or case study) as a descriptor (MacQueen et al., 1998). Although the interview protocol supplied an initial set of broad thematic areas, we also carried out an inductive analysis through line-by-line coding, which revealed unanticipated themes as well. To maximize project resources, one researcher first coded ten transcripts and then reviewed coded text with project team members. Discussions enabled the team to come to consensus on the types and number of codes; we then incorporated decisions into the coding for all interviews. Data saturation, or the point at which no new themes arose, occurred after the 16th interview. We continued with interviews to confirm that data saturation had been reached and to honor commitments to already scheduled interviews.

Thematic analyses integrated perspectives from community citizen science leaders and experts across the larger field of citizen science. This approach allowed us to understand how contextual elements could influence implementation and outcomes of community citizen science.

CHAPTER THREE

Case Study Descriptions

Each of our three case studies displays unique characteristics of community citizen science while also sharing commonalities with the other case studies across motivations, methods, and desired outcomes.

SkyTruth Pollution-Tracking Applications

SkyTruth is a nonprofit organization whose mission is to use remote sensing and digital mapping technologies to motivate the public to protect the environment (SkyTruth, undated a). John Amos, a geologist who became concerned about damaging human impacts to the environment, started SkyTruth in 2001. Since its inception, SkyTruth has grown into a small but wide-ranging operation. The organization acts as a watchdog, monitoring pollution and industrial activities through the use of satellite imagery and geospatial data. Results are used for multiple purposes, including advocacy, awareness-building, and collaborative research activities. SkyTruth pollution tracking applications represent the **collegial** form of citizen science, in that research is led by nontraditional organizations operating outside traditional research settings. Table 3.1 summarizes the governance structures and volunteer roles for each case study.

SkyTruth runs numerous projects addressing such issues as oil pollution, fracking, mining, and ocean vessel activity. Many projects incorporate volunteer engagement through crowdsourcing methods, such as soliciting public reports of environmental observations or providing image analysis tasks to identify industrial activities. Several SkyTruth projects were created in response to disaster events. For example, in 2010, in the midst of the *Deepwater Horizon* oil spill in the Gulf of Mexico, SkyTruth launched Oil Spill Tracker (SkyTruth, undated b). The tracker was a crowdsourcing platform that enabled public reporting of oil spill observations and impacts on the gulf coastline. More than 400 reports were uploaded to the platform documenting observations ranging from dead fish to business closures to human health effects. Oil Spill Tracker served as a resource for response and recovery groups on the ground, informed cleanup and recovery activities, and raised public awareness of potential impacts.

In late August 2017, Hurricane Harvey triggered widespread flooding throughout the highly industrialized city of Houston, Texas. Damage to chemical and polluting infrastructure resulted in public fears over a "toxic soup" of chemicals within floodwaters (Kaplan and Healy, 2017). Approximately one week after the storm, SkyTruth launched Spill Tracker, a map-based crowdsourcing tool for public reports on hazardous releases and other pollution incidents in Texas (SkyTruth, undated c). Around the same time, Hurricane Irma made landfall on the Leeward Islands and barreled through the Caribbean before reaching Florida and other south-

Table 3.1
Governance Structures of Case Study Citizen Science Models

Characteristic	Case Study		
	SkyTruth[a]	PRN	Surfrider Rincón
Model	Collegial	Collaborative	Collegial
Project initiator	Citizen	Academic	Citizen
Research lead	Citizen	Academic	Citizen
Citizen role			
Leadership[b]	Yes	No	Yes
Research[b]	Yes	Yes	Yes
Potential citizen research role			
Data collection	x		x
Data analysis		x	x
Data interpretation		x	x

[a] Community citizen science model.
[b] Citizen leadership and citizen researchers are not necessarily the same people.

ern U.S. states. Spill Tracker was extended to incorporate public reports from Florida and the Caribbean. The stated goal of Spill Tracker was to be "a useful resource for aiding the response and recovery efforts throughout the U.S. and the Caribbean" (Amos, 2017). Spill Tracker was used less extensively than Oil Spill Tracker but still provided lessons for project implementation and impacts. Our assessment included participant perspectives on both Spill Tracker and Oil Spill Tracker.

Planetary Response Network

The PRN is a collaborative endeavor that leverages the strengths of earth imaging organizations, academic institutions, platform developers, and disaster relief groups (Zastrow, 2016). The PRN is the culmination of efforts started in 2012 to determine how satellite imagery could be used to support ground-based disaster relief efforts. Deployed through Zooniverse, a platform for hosting crowdsourcing projects, the PRN relies on a robust partnership network that provides resources (e.g., satellite images) and infrastructure (e.g., platforms) needed to run virtual operations to inform on-the-ground relief efforts. It also relies on a network of volunteers to analyze satellite imagery of disaster-affected areas and perform damage assessments. Crowdsourcing data are combined with machine learning algorithms to accelerate analyses and improve the quality of volunteer classifications. The PRN is activated in response to specific requests by relief agencies to aid in disaster response and recovery.

The PRN's first official, public deployment was in 2016 in response to a 7.8-magnitude earthquake in Ecuador (Zooniverse, undated a). At the request of Rescue Global, a nonprofit disaster relief organization, the PRN was deployed to identify damaged areas and passable roads. More than 2,000 volunteers joined the effort and performed more than 75,000 image classifications. In just 12 hours, volunteers analyzed almost 25,000 square kilometers of satellite imagery. The PRN represents a **collaborative** form of citizen science, in that research is led by professional scientists or traditional research organizations, with citizens engaged in meaningful scientific tasks.

In the fall of 2017, multiple Caribbean island nations were struck by two hurricanes: Hurricane Irma in early September and Maria just a couple of weeks later. In response, Rescue Global requested help in assessing damage and identifying priority areas for rebuilding efforts (Zooniverse, undated b). The PRN deployment for Hurricanes Irma and Maria was open from September 12 to October 5, 2017. More than 5,000 volunteers joined, providing 650,000 classifications of images covering eight island nations. Volunteers classified various types of damage, including flooded areas and impassable roads, and identified temporary settlements. Analyses were used to produce heat maps of damage that Rescue Global used to aid in resource allocation and surveillance decisions.

The Surfrider Foundation Rincón Chapter's Blue Water Task Force

The Surfrider Foundation is a nonprofit organization with a network of volunteers who carry out Surfrider's mission "to protect and enjoy the world's oceans, waves and beaches" (Surfrider Foundation, undated a). Surfrider was started in 1984 by surfers in Malibu, California, who were concerned about coastal development around surfing spots and the potential threats posed to environmental and human health. Surfrider has since grown significantly and now consists of more than 170 volunteer chapters and student clubs across the United States and Canada that work on local and national initiatives.

Surfrider runs numerous programs designed to protect marine ecosystems, including the Blue Water Task Force (BWTF), a volunteer-run water testing, education, and advocacy initiative. The BWTF was launched in 1990, and its objectives are to monitor beach water quality, raise public awareness of coastal pollution, and work with stakeholders to implement solutions (Surfrider Foundation, undated b). The BWTF typically concentrates testing efforts on beaches not adequately covered by government monitoring programs. It posts its sampling protocols online to ensure adherence to quality standards. In 2017, the BWTF consisted of 40 chapter-run labs that processed more than 6,000 samples from 446 sites.

The Surfrider chapter in Rincón, Puerto Rico, launched its BWTF program in 2007, performing weekly tests for bacteria at various sites and processing samples in its lab (Surfrider Foundation Rincón, PR, undated). Since 2015, Surfrider Rincón has reported performing more than 1,000 water quality tests per year (Surfrider Foundation, 2016, 2017, 2018). For 2016, the chapter accounted for 26 percent of the national BWTF program's activity (Surfrider Foundation, 2017). Partnerships have enabled expansion of testing, provision of educational and training opportunities, and collaborative research opportunities. In addition, the Rincón chapter provides information to the public on a regular basis, alerts authorities to problems, and makes data publicly accessible. The Rincón BWTF represents a **collegial** form of citizen science, in that research is led by nontraditional research organizations, with programs run by nonprofessional scientists.

On September 20, 2017, Hurricane Maria made landfall on Puerto Rico, disrupting access to clean water supplies. Four weeks poststorm, Surfrider Rincón's BWTF program resumed water quality testing at beaches and freshwater sources (Dias, 2017). The original goal of the Rincón BWTF post-Maria was to provide enterococcus counts for freshwater sites in the Rincón and Aguada areas. However, as new partnerships developed, the chapter widened its scope by expanding into interior and mountainous areas, testing for coliforms and *E. coli*, evaluating water filtration systems, monitoring public drinking water supplies, and evaluating water testing methods. Communities used these results to identify unsafe household water sources, efficiently allocate filtration systems, and verify safe water supplies for the public.

CHAPTER FOUR
Thematic Analyses

Through thematic analyses of the case studies, we identified specific considerations for community citizen science implementation and impacts, while coded interviews with general citizen science experts provided a greater understanding of how the cultures and institutions that make up the larger field of citizen science might shape and react to the development of community citizen science. In this chapter, we summarize the three major themes arising from our interviews. Where relevant, we indicate when certain perspectives were voiced primarily by general experts or case study experts. At the end of the chapter, we summarize all the themes in Table 4.2.

Community Citizen Science and Traditional Scientific Institutions Reflect Different Cultures

Concerns About the Quality and Rigor of Community Citizen Science Arise Mainly from Traditional Scientific Institutions

The majority of interviewees characterized citizen science as a field encompassing multiple objectives. Examples included online gaming applications, environmental monitoring, disaster response research, and educational activities. In addition, interviewees noted differences in scales of activities, ranging from amassing large numbers of observations (i.e., crowdsourcing) to local applications in which smaller numbers engage in projects of interest.

Although there was agreement regarding citizen science as a "big tent," a few interviewees described a tension between "traditional" science led by professionals and "grassroots" science. Interviewees articulated the latter as science that citizens or communities "own," with *ownership* meaning anything from control over research direction and decisions to actual ownership of data. According to one interviewee, "to me, it's less about harnessing the public to help scientists advance their goals than it is about bringing the methods and tools of science to communities and ordinary people who might have very different agendas from [those of] traditional scientists." Another interviewee emphasized the value of focusing on citizenship in addition to science, saying, "Citizenship is key. The science part of it—we know what science is about. Citizen science is about owning—owning your opportunity."

However, some interviewees expressed concern that the rise of grassroots science might open the door for rigorous scientific methods to take a back seat to the achievement of community goals. To guard against this concern, a few general expert interviewees stated that the one unifying element of all citizen science, regardless of model, objectives, or who is in control, should be the production of "good science." As stated by one general expert interviewee,

13

"There's a huge range of things that people want to get out of these projects . . . . The science is the one thing that is truly in common . . . ." Another interviewee further noted, "I think that the goal for citizen science should be good science. That's all it really comes down to. It should be good, scientifically reliable."

Echoing the idea of performing good science, most case study interviewees emphasized the need to adhere to established scientific standards because they believed that community citizen science would not be taken seriously otherwise. In the Rincón BWTF case, one interviewee stated,

> I always advise . . . volunteers . . . Get your program developed. Work out the kinks; establish that everyone knows exactly what they're doing. The data is all making sense. I never encourage . . . chapters to go out and . . . issue a press release the first time they get a high hit of bacteria . . . . I really encourage them to get the protocols worked out so that they know exactly what they're doing.

Case study interviewees representing SkyTruth noted academic researchers' use of their satellite imagery in peer-reviewed publications and the desire to produce data of sufficient quality to be used in rigorous scientific research applications. General expert interviewees who had greater familiarity with community citizen science also pushed back on the concern over "good science":

> . . . Because we already know that we're going into a situation where the data and information is going to be under further scrutiny, we are extra careful in ensuring that we're following the exact same processes that scientists with a capital S would follow. So . . . just because you're using data or information to advocate or to try to better understand the context in which you're living . . . doesn't mean that you're going to do bad science. And I think that's a very popular misconception that people have about [community] citizen science.

The Distinction Between Professional and Community Citizen Scientists Is Unclear

Interviewees characterized citizen scientists as individuals, motivated by different reasons, who engage in research activities driven by scientific experts or through grassroots organizing. A few interviewees, however, highlighted freedom of choice as a defining element of a citizen scientist. Citizen scientists are people who freely choose to engage in scientific activities. Therefore, a citizen scientist is not defined by credentials or educational attainment (or lack thereof); the distinction is not between expert and nonexpert, paid and unpaid, or even professional and nonprofessional. According to a general expert interviewee,

> Citizenship is a volitional act by a citizen, someone who chooses to act because they recognize their responsibility as a member of a community on which they depend. And their action is of their own volition in order to advance some understanding through techniques and tools, through collaborating with others.

According to this perspective, professional scientists are also citizen scientists if they choose to use their skills to address issues of interest or concern to themselves or their communities. This thinking expands on the idea of citizen science as "public participation" or "public engagement" in science, into a larger philosophy of conducting science in the service of citizenship:

> . . . [Citizen science is] not even so much about public participation . . . . I sort of think of it like the whole open-source software movement . . . . Because all these guys who contributed to the open-source software movement, they were employed by places like Microsoft and other companies that were making proprietary software, so they had a stable paycheck and a job . . . . And then on their weekends, in the evenings, they were building software together, and the open-source software movement was all about sort of sharing, you know, working together, sort of divesting your ego . . . . And I kind of think that that's one of the most important currents going on in citizen science right now . . . .

Most interviewees agreed that movements within citizen science are pushing for greater openness, democracy, and equity. One case study interviewee envisioned a future in which questions about who a citizen scientist is end up being moot:

> . . . The ideal is to then essentially remove the entire notion of this person as a [citizen] scientist . . . . If we're all within the same framework asking pointed questions and going about a systematic way to answer them in reproducible ways, then we all end up being scientists.

Community Citizen Science Might Operate Under a Different Research Paradigm from That of Traditional Science

Discussions about the quality of community citizen science and the assessment of who is and is not a proper scientist shed light on some of the cultural differences that exist between traditional scientific institutions and the community citizen science movement. General expert interviewees' perspectives tended to be informed by an *academic research model*, which we define as a model in which performance of scientific research for the pursuit of knowledge is a professional endeavor. Membership within the academic research community can be a function of educational attainment (degrees), research activity, and institutional affiliation. Under this perspective, it makes sense to form distinctions between citizen and traditional science through dichotomies, such as expert versus nonexpert, because most "experts" perform scientific research under an academic research model. However, it is unclear whether the academic research model is the right paradigm through which to evaluate and advance community citizen science.

We considered two of our case studies—SkyTruth and Rincón BWTF—to be **collegial** citizen science models, in which projects are led by organizations operating outside traditional research institutions. Collegial citizen science can represent a type of "citizen research model" that is distinguished not by the degree of professionalism of its practitioners but by such elements as research goals, translational activities, provision of technical services, and types of institutional involvement (e.g., community or advocacy groups). For example, the two collegial case study interviewees shared such elements as a focus on research for direct and applied action, support for direct advocacy, a "watchdog" role in promoting institutional accountability, and the performance of services in the public interest (e.g., water quality testing, technological capabilities). Academic and citizen research models share a common interest in performing "good science," but differences in culture, goals, activities, and actors could have large ramifications for resource needs, implementation activities, and the acceptance and use of research. Given the larger share of resources held by academic and professional institutions, an open question is whether the academic research model should find ways to share institutional knowledge, resources, and capacities with the participants in citizen research models or whether these models should maintain their separation and develop on separate paths.

Related questions arise when considering **cocreated** and, to some extent, **collaborative** citizen science and the challenges that emerge when academic and citizen research models combine. These are not new questions, in that participatory research approaches have a long history of documented challenges (Freeman et al., 2006; Strickland, 2006; Cargo and Mercer, 2008; Ross et al., 2010). Common areas of community partner concerns, such as inequity in funding lines, exclusion from meaningful decisionmaking, and poor translation and evaluation mechanisms, could be mitigated if parties were to adopt a citizen research model to perform collaborative work. However, given the wide disparities in resources between the academic and citizen research models, it is often impractical to adopt such an approach. Even with collaborative models used in case studies, such as that for the PRN, in which citizens are involved primarily in meaningful scientific tasks (but not leadership roles), a recurring theme was that traditional institutions might still need to change to truly promote the success of collaborative science. In some cases, this theme refers to the need for scientific and policymaking institutions to change their perceptions of citizen science—for instance, by acknowledging, valuing, and incorporating different types of expertise into scientific research:

> I think many times, engineers, scientists, doctors, people who have a Ph.D. behind their name are automatically seen as the primary expert in a room. But what that does is . . . really pushes aside the other expertise that can exist and should kind of coexist. That's everything from practical skills that you learned while working on an assembly line in a factory to local knowledge because you've lived in a place for 30-plus years.

In other cases, it refers to the need to change existing structures or processes that might inadvertently serve to disincentivize researchers from pursuing collaborative research or exclude community collaborators from institutional knowledge, resources, and benefits.

Community Citizen Science Can Creatively Leverage Resources to Address Project Implementation Challenges

Community Citizen Science Often Faces Many Challenges to Project Implementation

Implementation of citizen science projects in general appears to face the same kinds of challenges regardless of whether citizen scientists are leading activities or mainly just participating in research. These challenges include (1) scaling and replicating activities, (2) balancing the roles of professional and citizen scientists, (3) navigating competition and duplication within the field, (4) clearly communicating project mission and objectives, and (5) obtaining needed project resources. In addition, disaster-related projects can exhibit their own set of challenges.

Scaling and Replicating Activities

Although citizen science activities appear to be proliferating, many general expert interviewees expressed concerns about the ability (or lack thereof) to scale or replicate projects at larger levels. One interviewee remarked that project leadership should understand the political and social context in which their work is conducted:

> . . . When we're working on one very specific issue, you can't separate the political and social context from the work that you're doing. And many times, lessons that come out of that context will not replicate or they won't help you to scale, so, for us, it's always thinking about what piece of our model, what pieces of our model, can grow and can be reused . . . .

Balancing the Roles of Professional and Citizen Scientists

Most interviewees saw an important role for professional scientists in ensuring that projects adhere to sound scientific methods, generate quality data, and properly handle complex analyses. This role was considered important regardless of whether projects were led by professionals or by citizens. For any endeavor in which a scientific or policy impact was a goal, interviewees believed that professional scientist involvement was necessary to some extent. Interviewees varied, however, on the roles they saw for citizen scientists. Some interviewees stressed the importance of always providing opportunities to volunteers to engage more deeply in analyses or discussion, while others noted that, often, volunteers are not interested in doing rigorous analytical or conceptual work. A few interviewees felt that professional scientists should step back and support communities to take the lead on citizen science projects, stepping forward only when professional expertise is requested or becomes necessary to achieving goals.

Navigating Competition and Duplication Within the Field

Many interviewees noted how the rapid evolution of citizen science has led to an explosion of projects and technological developments. With little coordination across activities, interviewees noted, duplicative efforts would just proliferate. Although competition was seen as a way to encourage innovation, interviewees felt that, without some organizing structure to help publicize and guide efforts, the true potential of citizen science effects might not be met. As one case study interviewee remarked,

> . . . I feel like the Red Queen running as fast as possible just to keep up with what's possible . . . . There's . . . a tendency . . . to always be out on the leading edge, which means you're always somewhere low on the learning curve . . . . There's a risk to that . . . . With the churn and the technology, we're not going to reach our potential levels of engagement . . . .

Case study interviewees also expressed that inadequate coordination across citizen science projects might also have consequences for volunteer participation rates because of inefficient allocation of efforts. The proliferation of activities requesting volunteers to perform the same, or even slightly different, tasks for different citizen science projects might result in confusion or diffusion of efforts. These concerns loom particularly large in the context of disaster response and recovery. Partnerships (e.g., government, software developers, academics, communities) or a central clearinghouse were seen as key to ensuring that future resources would not be developed in a silo and that efforts could be coordinated for maximum benefit. As stated by the same case study interviewee,

> I don't think there should be one player to rule them all, but I think there could be a more collaborative association of the entities and organizations that are doing these things so that they are less likely to be tripping over each other . . . .

Clearly Communicating Project Mission and Objectives

As summarized by a general expert interviewee,

> . . . I would say that [project leaders] should be very clear and transparent about their research goals, what is happening to the data, the data flows, and then also especially the outcomes of their research projects, and communicate that information back to their volunteers.

Project communication was considered important, first to ensure that volunteer and project leader expectations are aligned. Interviewees noted that activities often fail because of a mismatch in expectations, in which poor communication leads volunteers to expect an activity to deliver outcomes that it might not be designed to achieve. As noted by a case study interviewee,

> . . . I personally think it's important to have concrete objectives as an outcome to motivate people to participate . . . . When we do have that . . . I think we're on firmer ground ethically to ask people to help, and we get a better response because they can see what they're building toward.

Secondly, external communications to stakeholders or research end users could help achieve action-oriented goals. Interviewees felt that including both volunteers and stakeholders as early as possible in discussions would increase the potential for activities to be mutually beneficial and impactful for all:

> . . . I think it's really important in most projects . . . to have a two-way dialogue between . . . the core research team . . . and the core contributors . . . . Similarly, if there is an intention to have a policy or management impact, those people should be brought in as part of that core team from the start, because I think one of the . . . best ways to get around the lack of bridging that data results–to–policy gap is bringing in the policy from the start. And we're not doing that.

Obtaining Needed Project Resources

Six types of resources for citizen science project implementation were raised consistently in general expert and case study interviews: funding, technology and infrastructure, partnerships, volunteer participation, organizational support, and capable leadership.

Funding

Most interviewees discussed the critical nature that funding plays in supporting citizen science activities, but some noted that the current funding environment for citizen science is sparse:

> Citizen science has a really big sustainability problem, and a lot of that is access to money, and a lot . . . is coming up with sustainable business models . . . . Diversifying funding streams and then also getting the involvement not just of a [principal investigator] who is funded by [a] grant but working with community organizations to scale impacts and ensure that projects are going to be continued beyond . . . that initial . . . grant.

Funding difficulties can be particularly problematic for rapidly evolving fields, such as citizen science, in which advances are being made not just in academia but at the grassroots and community levels as well. Interviewees noted the need for funding to support coordination of the field across the full range of possible actors (e.g., community, nonprofit, academia, government, industry) to keep pace with advances. A few interviewees brought up the potential for corporate partnerships to fill funding gaps but also noted that most groups currently involved with citizen science do not have much experience partnering with industry and would need to develop relationships.

Technology and Infrastructure

Technology and infrastructure were also considered key to citizen science activities, with interviewees describing the various resources used in their projects. Altogether, these resources fell into five categories: equipment and related infrastructure, data tools and infrastructure, activity hosting platforms, clearinghouse or resource repositories, and technical assistance and communication mechanisms. In addition to describing existing technologies, interviewees offered suggestions and recommendations for technology and infrastructure needs that would enhance citizen science activities (see Table 4.1).

Interviewees also noted the importance of designing projects with tool usability and project goals in mind. Citizen science projects often use existing resources because they are freely available, even though they might not fully fit with the intended purpose of an activity. When technology is not appropriately designed, the consequences could be poor user experiences

Table 4.1
Technology- and Infrastructure-Related Resource Needs

Category	Suggested Resource Need
Equipment and related infrastructure	• Small, easy-to-use, and robust sensors and tools for environmental monitoring • Accessible and understandable standard operating procedures for use of equipment • Lending libraries for citizen science equipment to enhance access • Connecting with maker or do-it-yourself groups to design applications for use in larger research endeavors • Distributed community networks for housing equipment or launching activities
Data tools and infrastructure	• A central repository for data, along with tools to facilitate quality assurance • Tools to support data quality checks and verification, such as data quality modules that fit into common content management systems • Processes or technologies to make data usable and sharable across citizen science platforms
Activity hosting platforms	• Investment into platforms designed to support the majority of possible types of citizen science projects • Continued development of existing platforms for finding and connecting people to projects, and facilitating data exchange and general professional development of the field • Tools that support management of the overall participant experience (e.g., social aspects, helping volunteers find projects, keeping track of communications)
Clearinghouse or resource repositories	• A library that houses training and learning protocols and guidance for enforcement of such protocols in projects • Repositories to share methods, evaluation tools, results, program design, and other materials that would support development of best practices
Technical assistance and communication mechanisms	• A review board or panel that would provide feedback to citizen scientists on study design, statistical analyses, and methods • Further development of technologies for volunteer self-training, remote training, distributed training, and related support and materials • An academy for training researchers to work with citizen scientists and manage projects

or a lack of accessibility that curtails participation by a broader population and compromises results:

> . . . One of the biggest infrastructure issues that I see is usability and human-centered design. Those capacities are just missing in whatever technology resources that many of the projects are able to wrangle to support themselves . . . . Besides [a few platforms], almost everybody is struggling to . . . make their projects work with a tool that's maybe a 75-percent match, or they're trying to cobble together whatever they can using organizational resources or free stuff to get what they need.

Partnerships

The majority of interviewees said that having the right partnerships is integral to project implementation. Beyond contributing to the success of citizen science projects, partnerships were also seen as a way to expand activity reach, scope, and impact. In the case of Rincón BWTF, partnerships with community organizations allowed volunteers to expand testing and outreach to inland rural villages. For the PRN, partnerships facilitated the integration of robust sets of satellite images that strengthened outputs from volunteer analysts. One general expert interviewee described partnerships that could be especially fruitful:

> I would like to see better communication, which facilitates collaboration . . . by different communities who are doing similar things. So the environmental justice people: I would like them to be able to . . . work more closely with the citizen science people to . . . have conversations on . . . what are the benefits and what are the pitfalls to using established standards for data collection versus building your own protocols from the ground up. I'd like to see citizen science working with human computation and artificial intelligence communities to . . . train algorithms in a more responsible way and to combine human and machine intelligence to get more research done. I would like to see citizen scientists working with maker and hacker communities so that there could be a parallel pipeline to building data-collection tools . . . . Wouldn't that be great if we had . . . networks so that this wasn't just . . . a one-off study but was a longer-term partnership between different groups with similar interests and complementary activities?

Although citizen science was seen as an inherently collaborative activity, questions remained about how to form the right partnerships and work together. Most general expert interviewees focused on community–academic collaborations, while case study interviewees discussed a wider range of partners that provided such resources as technical training or support, administrative infrastructure, outreach to affected populations, media and communication capabilities, and research translation. Despite the importance of partnerships, interviewees noted difficulties in building and maintaining trusted relationships because of such factors as misaligned goals and expectations and administrative issues (e.g., conflicts over intellectual property rights, political turf wars, credit sharing, funding equity). As summarized by a general expert interviewee,

> . . . Any citizen science project worth its salt has, at the heart of it, a deep, complex collaboration across many professions. But we don't . . . pay enough attention to figure out how to identify the right kind of partners. How do you cultivate the relationships that you need in order to set the foundation for a collaboration? . . . But then also how do you see

that collaboration through time? Those are really important . . . skills that not many people have, and there's not a . . . lot of . . . conversation in the community about how. What are the skill sets involved, and how do you cultivate them and how do you practice them . . . ?

Volunteer Participation

The importance of volunteer participation and the challenges associated with motivating and retaining volunteers were frequently discussed topics. Interviewees described a common pattern in which a small proportion of highly committed volunteers (estimates of 5 to 20 percent) collect or analyze the vast majority of data and engage deeply in different research activities. Interviewees felt that work was needed to determine how best to meet the needs of high-performing volunteers, increase engagement of less committed volunteers, and ensure diversity in the volunteer force. Four factors potentially affecting engagement and participation were emphasized across interviewees.

First, **access to citizen science activities** can influence engagement. Project access can be affected by such issues as access to technology (e.g., Internet connection), the ability to travel to project sites, availability to volunteer, financial costs of participation, and awareness of opportunities. In addition, access can be influenced by personal characteristics, such as having certain skill sets or, as one interviewee remarked, "people simply [feeling] like they're . . . not capable of collecting data or participating in the scientific process." Access might be particularly relevant for underrepresented groups, such as communities of color or people of lower socioeconomic status, whom barriers might affect to a greater degree.

To counter access barriers, interviewees described employing different techniques, such as (1) providing plain-language and transparent communications that reflect the comfort level and expertise of the volunteer community, (2) ensuring that equipment is easy to use and understand, and (3) creating a welcoming community through accessible user interface designs or integration of collaborative processes between project leaders and volunteers. One general expert interviewee described going out to community spaces to introduce citizen science to underrepresented youth and listen to their interests: "The youth leadership council . . . [helps] us understand what the teens are interested in, where their starting points are, where the hooks are, how to meet them where their interests are, and then bring them into participation in citizen science . . . ."

Second, **aligning project objectives with volunteer motivations** was believed to help with engagement, but research is needed to determine how to identify and tap into different volunteer motivations. As stated by a general expert interviewee,

> . . . I think . . . more-targeted work . . . needs to be done to understand why volunteers want to become engaged and the different profiles of volunteers that are . . . out there and the more-specific features of a citizen science project that makes them come back and participate over time and inspire their loyalty . . . .

To this latter point on volunteer motivation, interviewees felt that it was important to figure out the incentives that would result in optimal participation and how to "create the . . . feedback loops that really engage people . . . ."

Third, **ensuring that volunteers feel respected** was considered important for engagement. Multiple interviewees brought up the need to not "waste" volunteers' time and to give back in ways that volunteers would appreciate. As noted by one general expert interviewee,

> . . . One kind of interaction . . . has to do with valuing the participation of the citizens and setting up ways [in which] the scientists can acknowledge and respond to the work of the participants. And it's become pretty clear . . . just looking across the broader world of citizen science, that, if participants don't feel recognized, acknowledged, and appreciated, then they won't last in a particular project . . . .

One case study interviewee described volunteer efforts as a gift:

> . . . . I think we have to be really careful and thoughtful about how we ask for [volunteer time] and then what we do when people give us that great gift . . . in terms of not abusing the crowd and making them think . . . all you want is free labor . . . and you're not going to give anything back.

Fourth, case study interviewees identified **project exposure** as a key factor influencing volunteer participation. Interviewees described inadequate recruitment and engagement in cases in which projects were not adequately publicized. To increase exposure, management of citizen science projects might entail outreach to media and community networks that can amplify messages and raise awareness of organizations, their work, and the need for volunteers.

Organizational Support

For collaborative models of citizen science, general expert interviewees mentioned that having a supportive organizational and leadership culture was an important factor affecting whether citizen science activities could flourish within an organization. A supportive culture was considered a key determinant of whether financial, administrative, physical, or human capital resources would be allocated toward activities, thus determining the sustainability of citizen science efforts and, to some degree, their success. One interviewee described a specific experience:

> . . . We had a big citizen science project . . . and the support of our leaders was critical to . . . getting the whole organization on board . . . . And the same thing is happening now . . . . One of our . . . projects . . . doesn't need to focus on citizen science. Well, the team working on that project has actually chosen to focus on citizen science, and I think that's an example of how . . . enthusiasm at the highest levels of the organization for citizen science has been an influence . . . .

Capable Leadership

All case study respondents mentioned the presence of either one person or a small group of people, including project leaders and citizen scientist volunteers, who were thoroughly committed to establishing programs that would meet their intended goals. According to one interviewee,

> . . . [In] any successful volunteer program . . . there's always at least one strong leader . . . . And it's awesome if that one strong leader has lots of volunteers who are committed and

dependable to participate and take the load off that one strong leader. But there's always someone who's really leading the charge . . . .

In addition to taking charge, interviewees said, successful leaders were ones who demonstrated flexibility, adaptability, and creativity in moving projects forward. Following Hurricane Maria, Rincón BWTF leadership demonstrated those traits, finding solutions to major challenges and leveraging community resources to restart water testing capabilities. In developing the PRN, leadership brought together the right skill sets and took inspiration from different fields to create a new resource for supporting disaster relief operations. Similarly, in developing SkyTruth pollution tracking applications, leadership drew creatively from existing applications for inspiration and to adapt technical capabilities for SkyTruth's specific use.

In cases in which volunteer participation is low, however, problems can arise if volunteer leaders shoulder too much of a burden to keep a program running. As noted by an interviewee,

> From time to time, when a volunteer-run program is not able to attract enough volunteers and things rest on one shoulder alone, if that volunteer [leader] moves or burns out or has a health issue, sometimes the programs can crash. There isn't anybody there to pick it back up again.

Disaster Response and Recovery Illustrate How Community Citizen Science Can Leverage Creative Solutions to Address Implementation Challenges

Disaster-related projects can exhibit their own set of challenges related to such issues as recruiting volunteers and maintaining engagement amid competing priorities and concerns, ensuring volunteer safety, and coordinating with partners and volunteers in situations requiring quick turnaround times. Interviewees discussed how disaster type could influence the challenges that arise. For example, SkyTruth's Oil Spill Tracker released in response to the *Deepwater Horizon* oil spill was considered successful, while a similar application implemented after Hurricane Harvey did not garner the same level of engagement. According to a SkyTruth case study representative,

> . . . . I think the BP spill is a different kind of disaster . . . For all the citizens and residents along the coast, there was nothing they had to deal with in terms of . . . taking care of their house and their families' immediate safety. So people . . . were coming down to the beach to see what's going on. And . . . that was a very different kind of disaster because people were not physically jeopardized in an immediate sense by what was happening . . . .

Interviewees also discussed how disasters can cause existing citizen science activities to adapt, change, and grow. For example, organizations that, like SkyTruth, monitor a variety of activities might find research endeavors suddenly expanding if a problem is discovered. This uncertainty complicates project implementation planning in such areas as partnership development, in that organizations might be unable to anticipate resource needs. Case study interviewees discussed the importance of establishing an organizational reputation in the community and developing personal relationships with influential community or organizational actors regardless of immediate project needs. These avenues could enhance perceptions of research credibility and open doors for future partnership development. The experience of the Surfrider Rincón BWTF was a prime example of the unpredictability of disaster-related citizen science and of the flexibility and creativity that low-resourced groups with strong community ties can exhibit.

After Hurricane Maria, the Surfrider Rincón BWTF expanded its water quality program into new geographic areas and increased the types of contaminants tested. The program also took on new responsibilities, intervening in the community based on research results. According to interviewees, Surfrider Rincón's success in program expansion could not have occurred without community partnerships, and it was important to understand the skill mix that might already have existed within a community and how to tap into those resources. As relayed by interviewees,

> . . . The storm hit . . . . There's no electricity anywhere, most people are without running water . . . . And you need electricity to run the normal lab . . . . And because . . . the Rincón chapter had established a reputation for [its] program before the storm hit, they were . . . able to reach out . . . and . . . get our lab back up and running. And so [we] went to this health center . . . . They had . . . a generator . . . . So [we were] able to get . . . equipment there . . . . That partnership was huge. We would have not been able to do it without that partnership. And the other partnership was with . . . [the Rincón Beer Company] relief center, which was set up as a place to . . . accept donations and goods . . . . So [the chapter] formed a partnership with this relief center so that [the national office] could send . . . the supplies . . . needed to do the testing . . . . Those were huge partnerships that allowed the Surfrider chapter to . . . get our program up and running so quickly after the storm.

The Rincón partnerships brought additional benefits. For example, having the lab within a health center created an impression within the community that the Rincón BWTF was addressing a health threat, not simply collecting data. The partnership with the relief center led to the emergence of a "community hub" for communicating sampling results and for recruiting and training volunteers. As noted by one case study interviewee,

> . . . They did a really great job of just connecting the dots with all the resources that the community had [and] just having . . . the relationships that they were able to build within the community, a lot of people really started to see the Rincón chapter's value, the Blue Water Task Force's value down there.

In some ways, case study interviewees representing the BWTF felt, the disaster helped to showcase the value of the organization's activities and gave the program higher visibility.

It is unknown whether the community partnerships would have developed as quickly if the Rincón BWTF had not already had existing relationships and a strong reputation on which to draw. For community citizen science groups, project implementation could depend on leveraging community resources and integrating new partner types that might not already think of themselves as having a role to play in scientific research. These types of partnerships could help offset the limitations in resources under which projects often operate and could be particularly useful in disaster settings, such as in Puerto Rico, if traditional scientific infrastructure is disrupted. Although lacking access to the resources and infrastructure provided by traditional scientific (e.g., academic) settings, collegial citizen science models might be better positioned to leverage local community resources and create a research-oriented community culture. Whether the resourcefulness exhibited by our case studies is indicative of non–disaster-related community citizen science efforts is unknown but an important area to explore for the larger field.

Conveying Credibility and Promoting Community Citizen Science Research Will Be Important for Meeting Policy-Related Objectives

More Than with Other Citizen Science Research Models, Policy-Related Benefits Can Motivate Community Citizen Science Research

Across the field of citizen science as a whole, interviewees discussed the potential for many scientific and social benefits, which often informed some of the underlying motivations for citizen science research. First, interviewees discussed how **citizen science, regardless of model, could enhance scientific research processes and outcomes**. In reference to the PRN, one case study interviewee noted, ". . . From a purely mechanical, data product–driven side of it, we literally need them [citizen scientists]. We can't do this without them." A SkyTruth case study representative talked about "creating data as an outcome. Data that aren't available any other way . . . the ability to create [and] extract unique data out of imagery to enable and inform scientific research—that's a great outcome." General expert interviewees described the potential cost and time savings that could be achieved for research that is dependent on collection or analysis of large volumes of data. One participant noted that, for these types of research projects, citizen science is a "tool for unlocking data that would otherwise remain not searchable, not discoverable."

Second, **citizen science could enhance educational opportunities and scientific literacy** and **could build problem-solving abilities and self-efficacy** for both volunteers and the larger public. An educated and literate population confident in its abilities to tackle problems was thought to lead to other benefits, such as better evaluation of scientific evidence by the public and more-meaningful engagement in decisionmaking. As stated by a Rincón BWTF interviewee,

> . . . A definite goal is training and demonstrating to the public how science is actually "done." With hands-on experience of field work and data collection, lab work, analysis of results, statistics, and drawing conclusions, creating hypotheses and designing experiments, something that is generally regarded as too complicated and mysterious for the layperson becomes comprehensible, to a large degree. At times, this awareness can show a community confronting a problem when . . . expert opinion and resources is absolutely required and that, at other times, it is within its own capacity and purview.

Third, **citizen science could increase public engagement in scientific and civic endeavors**. Multiple participants used the word "empowering" to describe citizen science activities, in that citizen science empowers communities to use scientific methods to solve local problems. In so doing, communities become less reliant on others to gather, interpret, and communicate information. In reference to the Rincón BWTF efforts after Hurricane Maria, one interviewee noted,

> . . . A really great impact has been . . . empowering communities and people to take care of their own business, to take charge of their own public health, and to take care of themselves . . . . It's not just licking your wounds. It's like, "let's do something."

Another Rincón BWTF interviewee elaborated on the larger community benefits that could result from such actions:

> Aside from the actual data collected, a definite benefit is developing a sense of community empowerment . . . . Ultimately, as the experience of Hurricane Maria has demonstrated, a community that already has developed some degree of technical [capacity] and management [or] organizing experience tends to suffer less under unexpected or disastrous conditions, is able to recover faster, and is in better position to effectively help neighboring areas to recover. That, in itself, is sufficient justification to promote these type of efforts further.

Finally, **citizen science could build or strengthen community networks**. Interviewees described how communities that formed in virtual or real-world settings allowed citizen scientists to engage with researchers and peers for social and problem-solving purposes. According to a PRN interviewee,

> People are mobilizing; people are inspiring; people are teaching each other how to do this; and that, to me, is amazing because . . . you have retired scientists, you have people in wheelchairs, people that are confined to beds, you have . . . middle school students, all going to the same place and they're teaching each other.

Although both general expert and case study interviewees mentioned scientific and societal benefits, general expert interviewees tended to describe project purpose in terms of achieving scientific or educational benefits, while case study interviewees characterized their work as a reaction to scientific or policymaking institutions not performing (or being unable to perform) real-world responsibilities. Therefore, the benefits that case study interviewees described were associated with achievement of policy goals or informing community actions. In the rest of this section, we list these types of benefits.

First, **citizen science could augment or extend the research or data-collection functions of existing scientific or policymaking institutions**. For example, in the case of the PRN, the rapid provision of damage assessments or "ground-based intelligence" to relief agencies in the field is currently filling a gap in disaster response agency capabilities. In addition, as case study interviewees said about the Surfrider BWTF program,

> . . . The agencies that monitor our beaches are not able to provide 100-percent coverage of our beaches year-round. The beach water quality monitoring programs that are run generally by county department of health programs or state environmental agencies—they don't have enough resources or funding or staff to cover all of their beaches, so they tend to focus on the most popular beaches or where they know there's more of a risk or pollution source. So that could leave a lot of beaches without water quality information . . . . And so that's how the [BWTF] program started, really to complement and provide more information than the county and state agencies could provide on their own, on where it's safe to swim and surf and where there are problems.

Case study interviewees valued collaborative endeavors wherever possible, noting,

> . . . Maybe there is an opportunity here for . . . government–public partnership in the adoption and application of new tools and approaches to societal concerns. And I could see how,

in that case, you're not asking society to fill a gap; you are asking society to go above and beyond . . . .

Second, **citizen science could supplant research functions that institutions do not or cannot perform**. When we asked case study interviewees about the factors that motivated them, many reported feeling that they had an obligation. In reference to the Rincón BWTF and the need for clean water post–Hurricane Maria, one interviewee remarked,

> It became very obvious very soon that public health agencies were not capable of meeting this absolute need, nor were federal agencies responding adequately (especially in more-remote mountain areas). So, since we had the equipment, materials, and experience to conduct basic assessments, we really had no choice . . . .

Third, **citizen science could monitor institutional activities to promote accountability**. In reference to a government crowdsourcing platform developed for damage reporting during a Colorado flood event, SkyTruth observed that

> . . . There still might be a role for us to play if the government launches an initiative that we feel is not sufficiently transparent, that they're collecting data and reports but . . . not making them available. Or they're filtering in a way that we think is a disservice to the public's right to know what's happening. So, in those situations, we may still find that there's a role for civil society to play . . . .

Fourth, **citizen science could raise awareness, provide information, and aid in risk communication efforts**. The Rincón BWTF used water quality results to inform the public and others about the safety of the drinking water supply. In addition to providing information, citizen science could be a tool for countering misinformation. In the case of SkyTruth's work during the *Deepwater Horizon* oil spill,

> . . . One of the motivations for Spill Tracker was [to] ground truth. Here's what people are really observing on the Gulf Coast . . . . I wanted people to realize how bad the actual situation was . . . . And I didn't want them getting distracted by . . . made-up stuff . . . . And that was a failure of government, basically, to fill the role as a trusted provider of robust information on a timely basis . . . . And hopefully we'll be better at it next time around, because I think the rumors hurt people almost as much as the reality in situations like that.

Although case study interviewees described many policy-related benefits, we do not know whether this focus would remain for community citizen science projects with applications beyond disaster preparedness. Disaster-related community citizen science might be particularly attuned to policy-relevant research applications because of the applied nature of disaster events and the need for research to directly inform response or recovery efforts. However, case study interviewees hinted at the promise of community citizen science for policy-relevant work apart from disaster settings:

> I think citizens can . . . provide valuable information to . . . academia or government to . . . help them focus on and understand better local conditions so that they can determine . . . Where should we be looking for problems? Where can solutions be best implemented to solve problems? . . . . I think . . . citizen volunteers, in basically any field . . . have a lot

of valuable information to share with government on where their resources can be best spent—where their attention and focus should be.

There Are Many Avenues to Help Convey Credibility of Community Citizen Science and Promote Research Uses

Given the relevance of disaster-related community citizen science for informing preparedness policies and actions, as well as the current level of institutional acceptance across entities, such as government, academia, and the private sector, conveying research credibility was considered highly important yet difficult. Although institutional acceptance is growing (Citizen Science Association, undated; CitizenScience.gov, undated), many interviewees pointed to a long road ahead for all forms of citizen science to be accepted, understood, and valued within the scientific community. According to one general expert interviewee,

> . . . We found that some of the biggest . . . opposition [to a public data-collection event is] coming from the scientists because [it's] . . . not the way things are done . . . . the idea is that the public can't do that, the public shouldn't be doing this, or it's just not the way it's done.

Overall, given the applied nature of community citizen science and the need to engage with stakeholders to achieve the field's goals, the larger scientific, policy, and community arenas' perceptions can greatly influence the development of the field. To this end, interviewees discussed ways to enhance credibility and promote research by addressing issues related to data quality, appropriateness of community citizen science research, citizen scientist motivations, research communication, volunteer training, and project evaluation.

Address Data Quality

Although many interviewees did not report feeling that poor quality data was a problem, they noted that, as a whole, citizen science receives "extraordinary scrutiny" and therefore must take into account how perceptions of data quality can hamper research uses and overall growth of the field:

> . . . One of the challenges with the science side of citizen science is that . . . a lot of scientists . . . are concerned about the data quality issues associated with citizen science. And sometimes those concerns are valid, and other times they are more . . . because of an inherent bias in the scientists, a bias towards the professionalization of science and a concern that . . . citizens can't collect—that ordinary people can't collect high-quality data.

Interviewees felt that some tools and mechanisms could sufficiently ensure data quality if rigorously applied, such as developing technical partnerships, instituting quality control processes, engaging volunteer communities in quality control exercises, incorporating machine learning or statistical modeling techniques to enhance reliability, or developing statistical consultation or review programs. In addition, interpretation of data quality could be aided by following the principle of "fitness for use." As defined in a 2015 White House memorandum, fitness for use is "the degree to which a dataset is suitable for a particular application or purpose, encompassing factors such as data quality, scale, interoperability, cost, and data format" (Holdren, 2015, p. 2). This principle recognizes that multiple levels of data quality can arise from citizen science activities and that the uses of such data should depend on the quality of those data. Fitness for use is a valuable principle because, as one participant stated,

> . . . I think data quality is always going to be a trade-off because the promise of citizen science is getting a lot of people involved, and the more people you have involved, the less quality control you're able to have or the less training you are able to offer.

Determine the Appropriateness of Community Citizen Science Research

A few general expert interviewees reported feeling that some research questions might be better suited to community citizen science than to traditional science and vice versa. Interviewees remarked that, when policy change or some action is an end goal, there might be a need to consider scientific approaches apart from citizen science, particularly those well suited to employing expensive measurement instruments or complex statistical analyses, to ensure that the data returned are of high quality. For policy impacts, some interviewees believed that citizen science data could be a "good indicator," that "professionals need to be called in for something," or that "something needs further examination" and that, as a result, those data might be better suited to inform earlier stages of decisionmaking. Community citizen science might be particularly useful in applications, such as air or water quality investigations, in which hypotheses could be generated about causes or impacts. As a general expert interviewee noted,

> . . . There are all kinds of different ways you inform policy, some of which—in early warning systems, your data quality is not the biggest issue . . . . You [are] just bringing attention to something that people can then come and look at with much higher quality. And so . . . I tend to believe that the policy applications [for citizen science] are going to be more on the end of things that we can find potential issues or potential discoveries early.

One issue, however, is that communication is needed between traditional and community citizen scientists on policy matters to ensure that expectations are aligned: "If that was clear from the get-go, you wouldn't have people disappointed that their data samples are only used as the indicator of the canary in the coal mine."

Understand Motivations

Interviewees noted that a growing distrust of institutions, or doubt about institutional capabilities to perform their responsibilities, might be influencing how community citizen science—particularly collegial models—develop and ultimately interact with other societal elements (e.g., government, industry, academia). As one interviewee stated,

> . . . I think a lot of times, [there is] continued frustration that the systems which are supposed to be protecting the health of our populations are ineffective. And it's sometimes . . . the direct result of a bad employee . . . . Sometimes it really is because they don't have the resources and infrastructure that is necessary to really provide robust support for people.

Another case study interviewee remarked on the potential for societal divisions to take root:

> . . . As society's faith in institutions fades and faith in scientists and academia . . . is also weakening, I think that there's a desire for people to know where their data comes from . . . . It's like . . . who collected this data . . . . That's me, or my tribe of watershed-monitoring friends and neighbors. And so I think that this kind of grassroots environmentalism could

be hand in hand with kind of a grassroots science, where people only trust what they've helped build.

If distrust translates into this type of tribalism, community citizen science might develop insularly, which could affect its ability to inform policy or translate research into action. Case study interviewees as a whole did not reflect this attitude and believed that, by motivating citizens to take greater responsibility in scientific and civic life, public feelings of distrust could be channeled into something positive.

Communicate Research Effectively

When planning communication efforts, interviewees recommended, one should think about the perceptions stakeholders might have about community citizen science and what approaches might best convince them about the validity of the data. According to a general expert interviewee, the approach taken should "depend on the risks and claims to authority that are needed by the parties that are engaged in decisionmaking." The interviewee further elaborated,

> If I was to walk in with citizen science data to a public policy meeting, I probably wouldn't bring up the fact that it was citizen science data. I'd probably say, "I have 25,000 data points here that we've analyzed," and I would . . . go straight to . . . why the policy recommendation comes from an inference drawn on a robust and comprehensive set of data . . . . If I'm looking at something that is about democratic decisions, then I'd talk about it as an outcome of a democratic process where the public has voted because of the way [the public has] already collected the data and allowed us to understand the answer.

Transparency was also considered a key factor for research presentation. One general expert interviewee recommended providing a standard set of information about data so others would know how it could be used:

> . . . What I see . . . is something like nutrition labeling for data sets. As in . . . what are . . . the main characteristics that a data user would need to know to make effective use of these data? And things around ethics, [intellectual property], license stuff, any kind of permissions on the data . . . . And then . . . the actual attributes of the data that make [the data] useful to anybody else . . . . If we could give people an easy snapshot, I want to believe that . . . would help us get more usage of it.

Train Volunteers

Interviewees agreed that volunteer training matters greatly, especially in light of the extra scrutiny citizen science receives. Interviewees discussed the need for research on the effectiveness of materials and distribution formats to improve training outcomes. In addition, a few interviewees discussed the lack of guidance for community citizen scientists about how to translate their research to a larger audience for impact. One interviewee discussed how their organization actively works with volunteers through material development and workshops to provide them with an understanding of the context in which they are doing their work and what that means for how the research could be presented and used.

Evaluate Project Processes and Outcomes

Case study interviewees listed a variety of impacts resulting from their work. The PRN defined *success* as meeting the needs of disaster relief organizations that requested damage assessments.

The Rincón BWTF described success as providing educational and risk communications that were informed by research results and the ability to prioritize communities that needed water filters. Although Spill Tracker did not engage users, SkyTruth discussed how, during the *Deepwater Horizon* disaster, Oil Spill Tracker informed cleanup activities and how data from other projects were used in academic research efforts. Although all of these impacts are notable, overall, interviewees conveyed challenges in conducting systematic program evaluations, such as measuring and tracking data usage and policy or community impacts. Other outcomes that interviewees deemed important, such as achievement of social goods (e.g., volunteer engagement, problem-solving capacity, educational growth) and scientific advances, were similarly difficult to track unless a program devoted resources to surveying its volunteer communities. Without such knowledge, the ability to make a convincing case for the value of community citizen science research becomes hampered.

Evaluation becomes all the more important considering the difficulties that many participants described in motivating and sustaining volunteer engagement. One case study, however, the PRN, was extremely successful in its efforts, motivating thousands of volunteers to devote their time to performing damage assessments for disaster events. Factors for success of the PRN might include the short time frame of required engagement (days to weeks) but also volunteers' knowledge that their efforts would directly aid disaster relief organizations in carrying out their missions and potentially save lives. The close integration between project activities and end-user needs might have served to motivate volunteers. In the end, most interviewees conveyed that, without evaluations that could demonstrate community, academic, or policy impacts, robust volunteer engagement would be difficult to sustain, making realization of the many potential benefits of community citizen science difficult to achieve.

We conclude this chapter with a summary of themes found with both our experts and our case study groups. See Table 4.2.

Table 4.2
Summary of Themes Across General Expert and Case Study Groups

Theme	General Expert	Case Study
Community citizen science and traditional scientific institutions reflect different cultures.		
Concerns about the quality and rigor of community citizen science arise mainly from traditional scientific institutions.	x	
The distinction between professional and community citizen scientists is unclear.	x	
Community citizen science might operate under a different research paradigm from that of traditional science.		x
Community citizen science can creatively leverage resources to address project implementation challenges.		
Community citizen science can face many challenges to project implementation.		
Scaling and replicating activities		
Balancing the roles of professional and citizen scientists		
Navigating competition and duplication within the field		x
Clearly communicating project mission and objectives		xx
Obtaining needed project resources		

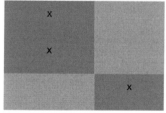
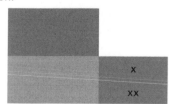

Table 4.2—Continued

Theme	General Expert	Case Study
Funding	xx	x
Technology and infrastructure	xx	xx
Partnerships	xx	xx
Volunteer participation	x	x
Organizational support		
Capable leadership		x
Disaster response and recovery illustrate how community citizen science can leverage creative solutions to address implementation challenges.		x

Conveying credibility and promoting community citizen science research will be important for meeting policy-related objectives.

Theme	General Expert	Case Study
More than with other citizen science research models, policy-related benefits can motivate community citizen science research.		
Citizen science enhances scientific research processes and outcomes.	xx	
Citizen science enhances educational opportunities and scientific literacy.	x	
Citizen science builds problem-solving abilities and self-efficacy.		
Citizen science increases public engagement in scientific and civic endeavors.	x	xx
Citizen science builds and strengthens community networks.		
Citizen science augments research functions of existing scientific or policy institutions.		x
Citizen science supplants research functions that institutions do not or cannot perform.		x
Citizen science monitors institutional activities to promote accountability.		x
Citizen science raises awareness, provides information, and aids in risk communication.		x

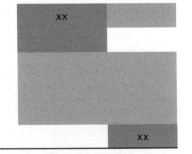

There are many avenues to help convey credibility of community citizen science and promote research uses.

Theme	General Expert	Case Study
Address data quality.	xx	
Determine the appropriateness of community citizen science research.		
Understand motivations.		
Communicate research effectively.		
Train volunteers.		
Evaluate project processes and outcomes.		xx

NOTE: No shading = This interview group did not discuss this theme. Any shading = This interview group discussed this theme. The group with the darker shading discussed the theme more than the other group did. x = This theme was among this group's moderately emphasized themes (the 50-percent most–frequently discussed themes). xx = This theme was among this group's highly emphasized themes (the 10-percent most–frequently discussed themes).

CHAPTER FIVE

Conclusions

Community citizen science is a growing and dynamic area filled with unique opportunities to engage the public in science and to enhance civic life. However, in addition to revealing the promise, our analysis revealed challenges facing the development of community citizen science, implementation of its activities, and translation and use of its research. Rather than being indicative of problems with the field, these challenges might reflect the rapid growth of community citizen science. As the nascent field evolves, these growing pains can serve to strengthen a movement that is searching for its place within, or beyond, existing scientific and policymaking institutions. In this chapter, we present reflections on lessons learned for community citizen science and future directions for research.

Community Citizen Science: Lessons Learned

We reviewed three community citizen science case studies for disaster applications: SkyTruth pollution tracking applications, Zooniverse's PRN, and the Surfrider Foundation's Rincón chapter BWTF. Two of the case studies operated under a **collegial** model of citizen science (SkyTruth and the Rincón BWTF) and one used a **collaborative** model (the PRN) (see Table 3.1 in Chapter Three). When we integrated case study interviews with general citizen science expert perspectives, we found that the larger field of citizen science provides important contextual elements that can inform the development of community citizen science. These elements include the challenges that citizen science as an overall field faces in gaining institutional acceptance; cultural differences between academic- and citizen-led initiatives that can influence structure, processes, and organization of community citizen science activities; and differences in the types and distribution of resources needed for traditional scientific projects and those for citizen science and community citizen science projects.

Although three is a small number, when we assess similarities and differences across the three case studies, we can still glean insights on potential factors contributing to project success. As illustrated in Table 1.1 in Chapter One, we defined *community citizen science* as either (1) citizens holding research leadership positions or (2) "meaningful" citizen engagement in aspects of the research process beyond data collection. Using the results of our analyses, in this section, we discuss how the organization of these two conditions could affect the implementation and outcomes of community citizen science for disaster-related applications.

Project or Research Leadership

Under the **collegial** model, SkyTruth programs and the Rincón BWTF were led by nontraditional research groups operating outside traditional scientific institutions. In contrast, under a **collaborative** model, the PRN was led by academic entities. Research leadership structure and organization can play a large role in conferring credibility of activities to funding entities, potential partners, and the larger public. Structure and organization can also be important in gaining access to institutional resources, such as money; labor; and administrative, technical, or knowledge infrastructures. The PRN might have been able to leverage resources provided by its academic credentials and the overall Zooniverse platform, such as a reputation for performing quality work and an existing volunteer network. These inherent traits might have helped to develop and sustain important partnerships, as well as enhance volunteer engagement. In contrast, although both SkyTruth and the Surfrider Foundation have strong reputations in their respective areas, partnerships with groups, such as academic or government institutions, were important for demonstrating scientific credentials and conveying legitimacy of research. For example, SkyTruth worked closely with academic partners during the *Deepwater Horizon* oil spill and on fracking impact studies (Asl et al., 2016; Casey et al., 2016; Rasmussen et al., 2016). Such research showed the value of SkyTruth's work in remote sensing for environmental observation. In the case of the oil spill, SkyTruth, in partnership with the Florida State University, examined satellite images and aerial survey data, which led them to challenge official reports on the amount of oil spilling in the Gulf of Mexico. Government estimates were subsequently revised upward, and the entire sequence of events was instrumental in bringing media and public attention to SkyTruth's mission and activities. Similarly, adding to its credibility, Surfrider Rincón's testing data are hosted on the Caribbean Regional Association for Coastal Ocean Observing (CARICOOS) website (Integrated Ocean Observing System, undated). CARICOOS is a regional coastal and ocean observing system that is funded by the National Oceanic and Atmospheric Administration and the goal of which is to provide tools and forecasts about coastal events (e.g., storms, water quality, sea levels) to inform actions for protecting and improving public safety, the environment, and the economy.

The Research Process

Our analysis identified resources that could be important for citizen science projects. We considered each resource's application to community citizen science. Table 5.1 summarizes recommendations that interviewees offered for addressing resource challenges.

Partnerships and Funding

As noted earlier in the report, community citizen science projects could be well suited to leverage existing community resources, including community partners. Given limited resources, community citizen science projects can require creativity and flexibility to develop relationships that could provide hard or soft resources, such as space, funds, equipment, social networks, project exposure and media attention, communication and dissemination channels, and organizational standing or credibility. Although partner-supplied resources can help partially offset funding needs, **collegial** community citizen science groups might be highly dependent on revenue sources, such as donations, foundation grants, membership dues, and merchandise sales. Unlike academic institutions, scientific research grants from foundations, private industry, or government entities likely make up a smaller amount of the organizational revenue stream in

Table 5.1
Interviewee Recommendations for Addressing Resource Challenges

Challenge	Recommendation
Partnerships and funding	• Diversify funding streams and reach out to nontraditional entities for help with resources (e.g., pursue corporate partnerships). • Develop "matchmaking" services to help projects find and form the right partnerships. • Establish organizational reputation within communities in which resources might be pursued. • Develop personal relationships with influential community or organizational actors regardless of immediate project needs.
Technology and infrastructure	• Leverage existing technologies and tools while taking into account design and usability for project goals and participants. • Develop clearinghouses or central mechanisms for greater communication, understanding, and collaboration between entities engaged in similar citizen science activities.
Volunteer participation	• Know the likely volunteer pool and make sure the project is accessible to its members (in terms of logistics, communications, technology, culture, and atmosphere). • Align project objectives with volunteer motivations, either through targeting recruitment to likely volunteers or by creating communication strategies that define a project's relevance for different groups of people. • Demonstrate to volunteers that they are valued by performing such actions as communicating often and transparently, providing feedback and updates, displaying gratitude, and channeling their efforts into impact. • Pursue media coverage for activities and issue communications in a wide range of venues to increase overall exposure and attention. • Ensure proper volunteer training by considering the best forms of training materials and training modalities.
Organizational support and leadership structures	• Wherever possible, widen the leadership circle, and promote or retain a robust volunteer base to allow for diffusion of responsibilities and to prevent burdens from falling on only a small number of people. • Identify and nurture highly committed volunteers to help with both project operations and creation of a larger culture of commitment to the project mission. • Create leadership training programs that integrate guidance on managing and overseeing both the scientific and social aspects of citizen science projects. • Regardless of citizen science model, design professional and citizen scientist roles according to project needs and intended outcomes. • Consider shared leadership structures whenever possible to promote citizen involvement in project governance.

collegial models. Therefore, a community citizen science project might need to directly couple its research work with a larger mission or cause that will help motivate financial contributions.

Technology and Infrastructure

Similar to partnerships, community citizen science groups can leverage existing publicly available technical resources to build the technological tools necessary for data collection and analysis. In addition, partnership development would be important to gain access to technological resources or augment or customize off-the-shelf resources. Limited resources might force community citizen science groups to explore cutting-edge or novel technologies that could help propel the field forward.

Volunteers

Community citizen science appears to suffer from the same challenges associated with volunteer recruitment and retention as other citizen science models. Both SkyTruth and Surfrider

Rincón noted difficulties in engaging a broad base of volunteers for data-collection efforts following their respective disaster events. The PRN, on the other hand, did not describe the same difficulties and was able to engage more than 2,000 volunteers in image classification tasks (Zooniverse, undated a, undated b). PRN research differed from the crowdsourcing and water quality testing activities carried out by SkyTruth and Surfrider Rincón. The PRN research tasks were time limited and highly focused in terms of goals and intended uses. In addition, the PRN had a specific partner on the ground that would be directly using research results. In contrast to research activities that were more open ended with a broader set of applications (e.g., awareness raising, risk communications), the built-in impact of PRN work coupled with its short-term nature might have proved to be highly motivating. Other important factors to consider are sources of volunteers and the overall accessibility of a project. PRN tasks could be undertaken by anyone with an Internet connection across the world; SkyTruth and Surfrider Rincón tasks, however, were more dependent on volunteers within disaster-affected areas who likely had competing priorities.

Organizational Support and Leadership

Across case study interviews, effective leadership stood out as an important factor for community citizen science success. In all three disaster-related projects, the presence of a strong leader who was resourceful, committed, and capable helped make programs work even when other resource elements were lacking (e.g., low volunteer participation rates) or technical difficulties arose. Effective leaders were responsible for the flexibility, adaptability, and creativity needed to identify community needs, form new partnerships, and find resources to establish or sustain a program following a disaster event. In more-established fields, leadership failure could potentially be offset by other actors or the presence of strong structures and processes that help programs function despite top-level difficulties. Given the small size of the community citizen science projects, it might have been the case that, without strong leadership, there would have been no disaster-related program. The dedication of the case study leaders made the case study programs work, but these programs need time to build and expand the leadership mind-set to other actors within the program. Although low volunteer participation rates put more pressure on leaders to both oversee projects and carry out project tasks, small leadership circles create the same pressures by placing full responsibility for project implementation on a single person.

Future Directions and Conclusion

Our work identified several research areas that will be important to pursue if the field of community citizen science is to move forward. First, more research is needed in general on the collegial model of citizen science. In their current work, scholars, such as Bonney and Shirk and their colleagues, have focused on understanding the design, outcomes, and impacts of three citizen science models: contributory, collaborative, and cocreated. These models likely represent the majority of current citizen science projects and are distinguished by academic leadership for research tasks (with varying levels of citizen involvement in either leadership or research processes). However, collegial models might be markedly different from the other citizen science models and can also exhibit variability in such elements as study designs, leadership organization, volunteer participation structures, and applications. Inventorying and assessing collegial citizen science models would shed light on future trends and help in understanding challenges that could inhibit growth of the field. Relatedly, research on the governance structures of community citizen science models is warranted—in particular, more explication on potential differences between academic versus citizen research models and implications for program development, activity outputs, outcomes, and future directions. If collegial models are developing and operating outside of traditional scientific institutions, it will be important to track the process to ensure harmony of efforts for elements that might be shared between the two cultures (e.g., conducting "good science"). Last, research is needed to explore community citizen science models (collaborative, cocreated, and collegial) in areas outside of disaster preparedness, response, and recovery. The lessons learned and conclusions we have drawn from our case studies are relevant primarily to disaster settings, but research should explore which practices might exhibit universality across all types of community citizen science projects.

Our study provides a foundation for understanding community citizen science, particularly within disaster-related applications. In the future, our work could be strengthened by assessing a greater range of community citizen science applications, including the cocreated citizen science model, and interviewing a broader set of stakeholders, including people directly affected by citizen science outputs and uses. These people include state, local, and federal government policymakers; industry stakeholders; media representatives; and community leaders and organizations. Their perspectives on how citizen science research could inform decision-making or actions within their organizations, as well as the challenges to its use, would be critical for guiding community citizen scientists on how best to conduct their research and present findings.

Overall, our work suggests that community citizen science is a complex phenomenon, with challenges and opportunities for both its practitioners and the larger societal structures its models might upend. Ultimately, regardless of whether science is carried out by professionals or amateurs, in academic labs or community backyards, what matters is a shared commitment by all to the conduct of "good science." With that shared commitment, community citizen science can be a mechanism for bringing different elements of society together, under a common goal of pursuing knowledge to improve the health and well-being of all.

References

Amos, John, "New Citizen Pollution Reporting Tool, Now Available for Hurricanes," SkyTruth, September 3, 2017. As of November 25, 2018:
https://www.skytruth.org/2017/09/skytruth-launches-citizen-pollution-reporting-tool-for-hurricane-harvey/

Asl, Samira Daneshgar, John Amos, Paul Woods, Oscar Garcia-Pineda, and Ian R. MacDonald, "Chronic, Anthropogenic Hydrocarbon Discharges in the Gulf of Mexico," *Deep Sea Research*, Part II: *Topical Studies in Oceanography*, Vol. 129, July 2016, pp. 187–195.

Bishop, Steven, "Citizen Science Is Stimulating a Wealth of Innovative Projects," *Scientific American*, October 1, 2014. As of November 25, 2018:
https://www.scientificamerican.com/article/citizen-science-is-stimulating-a-wealth-of-innovative-projects/

Bonney, Rick, Heidi Ballard, Rebecca Jordan, Ellen McCallie, Tina Phillips, Jennifer Shirk, and Candie C. Wilderman, *Public Participation in Scientific Research: Defining the Field and Assessing Its Potential for Informal Science Education*, Washington, D.C.: Center for Advancement of Informal Science Education, Inquiry Group Report, July 2009. As of January 7, 2019:
http://www.birds.cornell.edu/citscitoolkit/publications/CAISE-PPSR-report-2009.pdf

Browne, Janet, *Charles Darwin*, Vol. 1: *Voyaging*, Princeton, N.J.: Princeton University Press, 1995.

Cargo, Margaret, and Shawna L. Mercer, "The Value and Challenges of Participatory Research: Strengthening Its Practice," *Annual Review of Public Health*, Vol. 29, April 2008, pp. 325–350.

Casey, Joan A., David A. Savitz, Sara G. Rasmussen, Elizabeth L. Ogburn, Jonathan Pollak, Dione G. Mercer, and Brian S. Schwartz, "Unconventional Natural Gas Development and Birth Outcomes in Pennsylvania, USA," *Epidemiology*, Vol. 27, No. 2, March 2016, pp. 163–172.

Chari, Ramya, Luke J. Matthews, Marjory S. Blumenthal, Amanda F. Edelman, and Therese Jones, *The Promise of Community Citizen Science*, Santa Monica, Calif.: RAND Corporation, PE-256-RC, 2017. As of January 7, 2019:
https://www.rand.org/pubs/perspectives/PE256.html

Citizen Science Association, "About the Citizen Science Association," undated. As of November 25, 2018:
http://citizenscience.org/association/about/

CitizenScience.gov, "About CitizenScience.gov," undated. As of November 25, 2018:
https://www.citizenscience.gov/about/#

Coyne, Imelda T., "Sampling in Qualitative Research: Purposeful and Theoretical Sampling—Merging or Clear Boundaries?" *Journal of Advanced Nursing*, Vol. 26, No. 3, September 1997, pp. 623–630.

Dedoose, Dedoose version 8.0.35, Los Angeles, Calif.: SocioCultural Research Consultants, 2018. As of November 25, 2018:
http://www.dedoose.com

Dias, Mara, "Rincon BWTF: Community Empowerment Post–Hurricane Maria," Surfrider Foundation, December 4, 2017. As of November 25, 2018:
https://www.surfrider.org/coastal-blog/entry/rincon-bwtf-community-empowerment-post-hurricane-maria

Dickinson, Janis L., Jennifer Shirk, David Bonter, Rick Bonney, Rhiannon L. Crain, Jason Martin, Tina Phillips, and Karen Purcell, "The Current State of Citizen Science as a Tool for Ecological Research and Public Engagement," *Frontiers in Ecology and the Environment*, Vol. 10, No. 6, August 2012, pp. 291–297.

Eitzel, M. V., Jessica L. Cappadonna, Chris Santos-Lang, Ruth Ellen Duerr, Arika Virapongse, Sarah Elizabeth West, Christopher Conrad Maximillian Kyba, Anne Bowser, Caren Beth Cooper, Andrea Sforzi, Anya Nova Metcalfe, Edward S. Harris, Martin Thiel, Mordechai Haklay, Lesandro Ponciano, Joseph Roche, Luigi Ceccaroni, Fraser Mark Shilling, Daniel Dörler, Florian Heigl, Tim Kiessling, Brittany Y. Davis, and Qijun Jiang, "Citizen Science Terminology Matters: Exploring Key Terms," *Citizen Science: Theory and Practice*, Vol. 2, No. 1, 2017, pp. 1–20.

Freeman, Elmer R., Doug Brugge, Willie Mae Bennett-Bradley, Jonathan I. Levy, and Edna Rivera Carrasco, "Challenges of Conducting Community-Based Participatory Research in Boston's Neighborhoods to Reduce Disparities in Asthma," *Journal of Urban Health*, Vol. 83, No. 6, November 2006, pp. 1013–1021.

Haklay, Muki (Mordechai), *Citizen Science and Policy: A European Perspective*, Washington, D.C.: Woodrow Wilson International Center for Scholars, Science and Technology Innovation Program, Commons Lab, Case Study Series, Vol. 4, February 5, 2015. As of January 7, 2019:
https://www.wilsoncenter.org/publication/citizen-science-and-policy-european-perspective

Holdren, John P., assistant to the President for science and technology and director of the Office of Science and Technology Policy, Executive Office of the President, "Addressing Societal and Scientific Challenges Through Citizen Science and Crowdsourcing," memorandum to the heads of executive departments and agencies, September 30, 2015. As of January 7, 2019:
https://obamawhitehouse.archives.gov/sites/default/files/microsites/ostp/holdren_citizen_science_memo_092915_0.pdf

Hopkins, Graham W., and Robert P. Freckleton, "Declines in the Number of Amateur and Professional Taxonomists: Implications for Conservation," *Animal Conservation*, Vol. 5, No. 3, August 2002, pp. 245–249.

Integrated Ocean Observing System, "CARICOOS," homepage, undated. As of November 25, 2018:
https://www.caricoos.org/

Kaplan, Sheila, and Jack Healy, "Houston's Floodwaters Are Tainted, Testing Shows," *New York Times*, September 11, 2017. As of November 25, 2018:
https://www.nytimes.com/2017/09/11/health/houston-flood-contamination.html

Leavitt, Sarah A., "Gregor Mendel: The Father of Modern Genetics," *Deciphering the Genetic Code: Marshall Nirenberg*, Office of History, National Institutes of Health, U.S. Department of Health and Human Services, June 2010. As of November 25, 2018:
https://history.nih.gov/exhibits/nirenberg/hs1_mendel.htm

MacQueen, Kathleen M., Eleanor McLellan, Kelly Kay, and Bobby Milstein, "Codebook Development for Team-Based Qualitative Analysis," *Cultural Anthropology Methods*, Vol. 10, No. 2, June 1998, pp. 31–36.

Newman, Greg, Mark Chandler, Malin Clyde, Bridie McGreavy, Muki Haklay, Heidi L. Ballard, Steven A. Gray, Russell A. Scarpino, R. Hauptfeld, David T. Mellor, and John A. Gallo, "Leveraging the Power of Place in Citizen Science for Effective Conservation Decision Making," *Biological Conservation*, Vol. 208, April 2017, pp. 55–64.

Ottinger, Gwen, "Buckets of Resistance: Standards and the Effectiveness of Citizen Science," *Science, Technology, and Human Values*, Vol. 35, No. 2, June 2009, pp. 244–270.

Patton, Michael Quinn, *Qualitative Research and Evaluation Methods*, 3rd ed., Thousand Oaks, Calif.: Sage Publications, 2002.

Public Lab, "About Public Lab," undated. As of November 25, 2018:
https://publiclab.org/about

Rasmussen, Sara G., Elizabeth L. Ogburn, Meredith McCormack, Joan A. Casey, Karen Bandeen-Roche, Dione G. Mercer, and Brian S. Schwartz, "Association Between Unconventional Natural Gas Development in the Marcellus Shale and Asthma Exacerbations," *JAMA Internal Medicine*, Vol. 176, No. 9, September 2016, pp. 1334–1343.

Ross, Lainie Friedman, Allan Loup, Robert M. Nelson, Jeffrey R. Botkin, Rhonda Kost, George R. Smith Jr., and Sarah Gehlert, "The Challenges of Collaboration for Academic and Community Partners in a Research Partnership: Points to Consider," *Journal of Empirical Research on Human Research Ethics*, Vol. 5, No. 1, March 2010, pp. 19–31.

Shirk, Jennifer L., Heidi L. Ballard, Candie C. Wilderman, Tina Phillips, Andrea Wiggins, Rebecca Jordan, Ellen McCallie, Matthew Minarchek, Bruce V. Lewenstein, Marianne E. Krasny, and Rick Bonney, "Public Participation in Scientific Research: A Framework for Deliberate Design," *Ecology and Society*, Vol. 17, No. 2, 2012, art. 29. As of January 9, 2019:
https://www.ecologyandsociety.org/vol17/iss2/art29/

SkyTruth, "About SkyTruth," undated a. As November 25, 2018:
https://www.skytruth.org/about/

———, "Oil Spill Tracker," undated b. As of November 25, 2018:
https://www.skytruth.org/oil-spill-tracker/

———, "SkyTruth Spill Tracker," undated c. As of November 25, 2018:
https://skytruth.ushahidi.io/views/map

Smith, Elta, Sarah Parks, Salil Gunashekar, Catherine A. Lichten, Anna Knack, and Catriona Manville, *Open Science: The Citizen's Role and Contribution to Research*, Santa Monica, Calif.: RAND Corporation, PE-246-CI, 2017. As of January 9, 2019:
https://www.rand.org/pubs/perspectives/PE246.html

Strickland, C. June, "Challenges in Community-Based Participatory Research Implementation: Experiences in Cancer Prevention with Pacific Northwest American Indian Tribes," *Cancer Control*, Vol. 13, No. 3, July 2006, pp. 230–236.

Surfrider Foundation, "34 Years of Coastal Protection," undated a. As of November 25, 2018:
http://history.surfrider.org/

Surfrider Foundation, "Blue Water Task Force," undated b. As of November 25, 2018:
https://www.surfrider.org/blue-water-task-force

———, *Clean Water: Annual Report 2015*, San Clemente, Calif., c. April 2016. As of November 25, 2018:
http://publicfiles.surfrider.org/Clean-Water_Annual-Report-2015.pdf

———, *Clean Water Report: 2016*, San Clemente, Calif., c. April 2017. As of November 25, 2018:
http://publicfiles.surfrider.org/SF-Clean-Water-Annual-Report-2016.pdf

———, *Clean Water Report: 2017*, San Clemente, Calif., c. April 2018. As of November 25, 2018:
http://publicfiles.surfrider.org/Clean_Water/Clean-Water-Report_042018.pdf

Surfrider Foundation Rincón, PR, "BWTF Water Testing Program," undated. As of November 25, 2018:
https://rincon.surfrider.org/programs/bwtf-water-testing-program/

van Wyhe, John, "Alfred Russel Wallace: A Biographical Sketch," *Wallace Online*, undated. As of November 25, 2018:
http://wallace-online.org/Wallace-Bio-Sketch_John_van_Wyhe.html

Wilderman, Candie C., "Models of Community Science: Design Lessons from the Field," in C. McEver, R. Bonney, J. Dickinson, S. Kelling, K. Rosenberg, and J. Shirk, eds., *Citizen Science Toolkit Conference, Cornell Lab of Ornithology, June 20–23, 2007*, proceedings, Ithaca, N.Y., 2007, pp. 83–97. As of November 25, 2018:
http://www.birds.cornell.edu/citscitoolkit/conference/toolkitconference/proceeding-pdfs/Full%20Proceedings.pdf

Zastro, Mark, "Citizen Scientists Aid Ecuador Earthquake Relief," *Nature News*, May 3, 2016. As of November 25, 2018:
https://www.nature.com/news/citizen-scientists-aid-ecuador-earthquake-relief-1.19861

Zooniverse, "Planetary Response Network and Rescue Global: Ecuador Earthquake 2016," undated a. As of November 25, 2018:
https://www.zooniverse.org/projects/vrooje/planetary-response-network-and-rescue-global-ecuador-earthquake-2016/about/research

———, "Planetary Response Network and Rescue Global: Caribbean Storms 2017," undated b. As of November 25, 2018:
https://www.zooniverse.org/projects/vrooje/
planetary-response-network-and-rescue-global-caribbean-storms-2017/about/research